W9-BSV-201

COUNTDOWN TO SPACE

SPACEWALKS

The Ultimate Adventure in Orbit

Gregory L. Vogt

Series Advisor:
John E. McLeaish
Chief, Public Information Office, retired,
NASA Johnson Space Center

Enslow Publishers, Inc.

40 Industrial Road   PO Box 38
Box 398   Aldershot
Berkeley Heights, NJ 07922 Hants GU12 6BP
USA   UK

http://www.enslow.com

Library of Congress Cataloging-in-Publication Data

Vogt, Gregory L.
    Spacewalks / the ultimate adventure in orbit / Gregory L. Vogt.
        p. cm. — (Countdown to space)
    Includes bibliographical references and index.
    Summary: Describes the training and preparation for spacewalking, the
hazards faced by astronauts, as well as the construction of spacesuits.
    ISBN 0-7660-1305-7
    1. Extravehicular activity (Manned space flight) Juvenile literature.
    [1. Extravehicular activity (Manned space flight) 2. Manned space flight.]
    I. Title. II. Series.
    TL1096.V64   2000
    629.45'84—dc21   99-37094
                                                              CIP
                                                               AC

Printed in the United States of America

10 9 8 7 6 5 4 3 2 1

To Our Readers: All Internet addresses in this book were active and appropriate
when we went to press. Any comments or suggestions can be sent by e-mail to
Comments@enslow.com or to the address on the back cover.

Photo Credits: Archives and Manuscripts Division of the Oklahoma
Historical Society, p. 17; National Aeronautics and Space Administration
(NASA), pp. 4, 7, 8, 11, 13, 15, 19, 20, 24, 25, 26, 29, 30, 33, 35, 36, 38,
39, 40.

Cover Illustrations: NASA (foreground); Raghvendra Sahai and John
Trauger (JPL), the WFPC2 science team, NASA, and AURA/STScI
(background).

*Cover Photo: Tamara Jernigan performs a spacewalk during construction of the
International Space Station.*

CONTENTS

Astronaut Kathryn Thornton performs a spacewalk to repair the Hubble Telescope. Thomas Akers joined her and can be seen in the lower right corner.

1

The Ultimate Adventure

Astronaut Kathryn C. Thornton was suspended in space. During her December 1993 spacewalk, her feet were clipped to the end of the space shuttle's fifty-foot-long robotic arm. Nearly 370 miles below was the African Sahara. It was nighttime over the shifting sand dunes, but soon the shuttle would move into sunlight. Her spacewalking partner, Thomas D. Akers, looked at the stars passing above. "It looks like flying in a T-38 [jet aircraft] at night with lights on the ground."[1]

In her hands, Thornton held a 350-pound solar array. A solar array converts sunlight into electricity. The array was bent and damaged. With daylight streaking over the nation of Somalia, Thornton released the array toward Earth. As the array slowly fell, Thornton and

Akers replaced the damaged array with an improved model.[2]

Later, at a press conference on Earth, Thornton narrated a video of the mission. "The solar array begins flapping and it looks like a giant soaring bird over the desert. It was an incredible sight. I was mesmerized."[3]

Kathryn Thornton and Tom Akers were working on the Hubble Space Telescope. They were spacewalking outside the space shuttle. A spacewalk is what astronauts do when they venture outside their spaceship. The National Aeronautics and Space Administration (NASA) calls spacewalks extravehicular activities, or EVAs. Because outer space is deadly, they had to wear special protective clothes called space suits.

The Hubble Space Telescope was launched by NASA in 1990. It was made to peer deeply into the universe and discover many new secrets. Unfortunately, an error was made in the telescope's construction. Its giant mirror was shaped incorrectly, and the telescope could not focus clearly.

During the three years the Hubble Space Telescope had been in space, other problems had cropped up. The solar arrays that made electricity from sunlight continually warped as the spacecraft orbited into and out of the Sun's heat. This problem was one of the reasons spacewalkers Thornton and Akers were working on Hubble.

The two astronauts were not the only ones working

The two large solar panels on the Hubble Telescope provide its power. After the spacewalkers repaired the telescope, it was ready to provide the world with amazing pictures of the universe.

on the ailing spacecraft. The entire crew of the space shuttle *Endeavour* was involved. The mission commander and pilot controlled *Endeavour* so that it could meet Hubble in orbit. Another astronaut operated the robotic arm and captured the telescope. Later, the arm was used to move the spacewalkers and to help them lift replacement parts the size of telephone booths. Mission Control in Houston commented on the task. "The [teamwork] between the crewmember operating the arm and the person that is out on the end of it . . . is a marvelous way to do business. This saves a tremendous amount of time."[4]

The day before Thornton's and Aker's spacewalk, two more astronauts put on their space suits. Jeffrey A. Hoffman and Story Musgrave started the repair mission by unpacking tools they needed for the job. They replaced several gyroscopes. Gyroscopes are heavy rotating wheels like giant spinning tops. They help keep

Jeffrey Hoffman and Story Musgrave, while attached to the robotic arm, work on Hubble during a spacewalk.

Hubble from tumbling, and they point the telescope in the right direction. After the installation, Mission Control radioed to the crew. *"Endeavour,* Houston. Not to get you spun up but we have six good gyros on the telescope."* Hoffman replied, "Fantastic!"*[5]

By the end of *Endeavour's* ten-day, nineteen-hour-long mission, the spacewalking teams of Thornton and Akers and Hoffman and Musgrave made a record total of five spacewalks.[6]

The spacewalkers also replaced scientific instruments to correct the telescope's focus problem. It was as if Hubble were given a pair of glasses to wear. Near the end of the mission Hoffman stated, "We've got a basically new telescope there. And it can be really exciting for the astronomical community and the whole world to see what Hubble can do with a new set of eyeballs."[7]

At nearly 12:30 A.M. on December 13, 1993, the space shuttle *Endeavour* glided to a landing at NASA's Kennedy Space Center in Florida.[8] The astronauts were exhausted but very excited. Their spacewalking mission had been a complete success! Astronomers were delighted with the finely focused images Hubble was getting of distant objects in space. Spacewalks helped make it possible.

2

The Hazards
of Space

Spacewalking is anything but routine, and it is very dangerous. Outer space is unlike anything most people have ever experienced. If a person went into outer space wearing everyday clothing, his or her life expectancy would be about thirty seconds![1]

The first thing that would happen is that all the air in the lungs would come rushing out. It would not be possible for a person to breathe, because there is no air to breathe. Space is a vacuum. In other words, it is a place that does not contain any matter, including air.

Earth's atmosphere, a collection of gases such as oxygen and nitrogen, is held near Earth's surface by Earth's gravity. The atmosphere surrounds Earth to a height of about one hundred miles. Most of the atmosphere is found close to the surface. It thins out the higher you go and is just about gone above one hundred miles.

When the space shuttle goes into space, it needs to get above Earth's atmosphere. Therefore, the shuttle travels into the vacuum of space. While inside the space shuttle, astronauts are protected from the vacuum by air held in the cabin of the spacecraft. Outside the shuttle, however, there is no air, which means there is no air pressure, either. On Earth, every minute of every day

Above Earth's atmosphere, there is no air pressure. Astronauts going on a spacewalk need their own air supply as well as protection from the environment of space.

people are squeezed by tons of air weighing down on their bodies. If they were exposed in space, the lack of air pressure would cause the air inside their lungs to rush out. They would begin to suffocate. The problems would not stop there. Without the pressure, the skin would begin to swell because gas bubbles would form in the liquids in the skin cells. The reaction is similar to taking the top off a bottle of soda. Dissolved gas bubbles in the beverage start bubbling out because the pressure has been released.

The eardrums would burst because the pressure inside the inner ear would no longer be balanced by outside pressure. Finally, bubbles would form in the bloodstream, blocking the vessels in the brain. This would make the vessels burst, causing death.[2]

It all sounds gruesome enough, but there is more. Without an atmosphere, the Sun's rays are extremely intense. The side of a person facing the Sun would be heated to as much as 350°F. The side facing away from the Sun would drop to –250°F, cooking and freezing a person at the same time.[3]

Impacts with small particles could occur, such as dust from comets, tiny bits of rock left over from the collisions of asteroids, and metal and paint fragments from exploded rocket boosters. Comets are like dirty snowballs that orbit the Sun and give off long, streaming dust-filled tails when the Sun warms them. Asteroids are small, moonlike bodies of rock and metal

Dust from comets poses another threat to spacewalking astronauts.

that orbit the Sun and occasionally collide with each other. When comet and asteroid particles pass near Earth, they are traveling very fast. They could hit a person at speeds of 44,600 miles per hour or more.[4] It would be similar to being shot with tiny bullets that blast out small craters in anything they hit. The metal and paint are usually not traveling as fast as the debris from space, but they can be equally deadly.

Clearly, astronauts planning to go on a spacewalk need special protection. They need a space suit that acts like an eggshell to hold in air pressure and pleasant temperatures. At the same time, the space suit needs to be flexible enough to permit the astronaut to move around. These requirements turned out to be quite a challenge.

3

Building a Space Suit

When astronauts first left the safety of their space capsules, little was known about how to design a space suit and make it comfortable to wear. The first spacewalker was Alexei Leonov, a cosmonaut from the former Soviet Union.[1] Leonov flew into space in a spaceship called the *Voskhod 2*. On March 2, 1965, he emerged through a hatch in the side of the spaceship. The spacewalk lasted only twenty-four minutes, and it nearly ended in disaster. Air pressure inside Leonov's suit caused the suit to expand like a balloon, making it difficult for him to bend. Leonov got stuck for a time in the airlock of the Voskhod when he tried to get back inside.[2]

American astronaut Edward White II became the

America's first spacewalker, Edward White, breathed oxygen that was supplied through the oxygen line in the tether.

second spacewalker when he ventured outside the *Gemini 4* space capsule three months later, on June 3, 1965. White found it difficult to move about in his space suit, and the cooling system was not working very well. Soon, he began perspiring heavily, and the inside of his space helmet fogged, making it very difficult to see. White remained outside the *Gemini 4* capsule for twenty minutes. To make sure he did not get too far from the capsule, White was held by a twenty-five-foot tether that contained an oxygen line for breathing.[3] He had an excellent time. "There was absolutely no sensation of falling and very little sensation of speed," White told fellow astronaut Gus Grissom after returning to Earth.[4]

Designers of early space suits were trying to build a protective suit to keep the astronauts alive in one of the deadliest environments imaginable. The designers had little experience and much to learn. They had to use the experiences of builders of other kinds of garments such as suits of armor, deep-sea diver suits, and stretchy underwear.

One of the first attempts at building a space-suit-like garment took place in 1934. Airplane pilot Wiley E. Post wanted to compete in airplane races. Post knew that airplanes can fly faster the higher they fly, because the air thins out. Post's problem was that the thin air could make him sick or even kill him if he flew too high for too long.

Post had the BFGoodrich rubber company build a pressure suit for him. It looked very much like a deep-sea diver's suit and even had a metal helmet with a small glass porthole. Post did not need a larger porthole because he was blind in one eye.[5]

When inflated, the suit was very stiff, and it ruptured during pressure tests. The second suit for Post was built in the shape of a pilot sitting at the controls of an airplane. This design reduced the amount of bending Post would have to do while flying.

Post ran into a problem when he tried to put the suit on. The Goodrich engineers had not noticed that Post had put on some extra weight since they had first measured him. The suit was too tight and he got stuck inside. To get him out, the engineers had to cut the suit off Post. The third suit fit, and Post safely flew up to an altitude of 50,000 feet. (For comparison, commercial jets normally fly at about 35,000 feet.)

Eventually, flying at high altitudes became quite

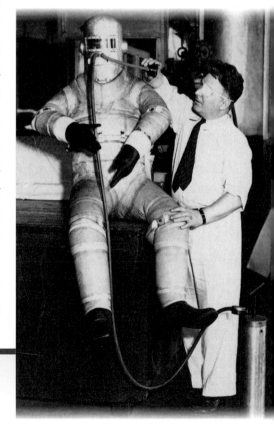

Wiley Post checks out his pressure suit, which enabled him to fly a plane at 50,000 feet.

common, and a variety of high-altitude suits were built. Some suits looked like medieval torture devices, and others made the wearer look like Hollywood movie aliens. One test suit was actually built for a Saint Bernard dog.[6]

With every new space mission, space suit engineers continued to improve space suits and made them more comfortable to wear and safer to use. Today, space suits are called extravehicular mobility units, or EMUs. They are worn by space shuttle astronauts when repairing satellites and assembling structures outside the space shuttle.

The EMU is a multilayer garment. Each layer serves a specific purpose. All together, the layers contain an atmosphere, regulate temperatures, and protect the spacewalker from the dangers of space.

The innermost layer of the EMU looks like a suit of long underwear that has plastic tubes woven through the fabric. It is called the liquid cooling and ventilation garment (LCVG). Cool water circulating through the tubes keeps the wearer cool. Attached to the outside of the LCVG are large tubes for circulating air through the suit. Above the LCVG is a double layer of nylon coated with rubber on the inside. This layer is like a rubber balloon that keeps air inside the suit. When oxygen is pumped into the suit, this layer fills up, but the nylon does not permit it to expand beyond a certain size. As more oxygen is pumped in, the oxygen starts to squeeze

*Story Musgrave is wearing a liquid cooling and ventilation garment.
Plastic tubes in the fabric carry cool water to keep him cool.*

in on the astronaut. This provides the pressure needed
to keep the astronaut's body fluids from boiling.

The trouble with pressure layers is that they get stiff
when inflated. Joints built into the suit at the ankles,
knees, waist, shoulders, elbows, and wrists solve the
problem. The joints are made of metal rings and cables
that pinch the suit so the wearer can bend more easily.

Above the pressure layer are several layers of
insulation material made of thin plastic. The plastic is
coated with aluminum to shield the astronaut from the
extreme temperatures in space. Finally, there is an outer
layer of fabric similar to the fabric used for bulletproof
vests. This layer protects the astronauts from collisions

with tiny high-speed micrometeoroids (dust from comets, asteroid fragments, and pieces from old spacecraft traveling at speeds of tens of miles per second).

The EMU is finished off with gloves for the hands, boots for the feet, and a helmet for the head. The helmet is a clear plastic dome like an upside-down goldfish bowl. The visor assembly, placed above the helmet, is

President Bill Clinton tries on a glove from the extravehicular mobility unit (EMU) space suit.

made up of several sliding visors to block out sunlight. A gold-coated face shield acts like a giant pair of sunglasses. Finally, a pair of small headlamps is attached to the outside.

A control panel worn on the chest has various knobs and switches for controlling the oxygen supply, temperature, and radio communications. A small number and letter display, similar to the narrow window on a calculator, tells the spacewalker how the various suit systems are working.

A large white backpacklike box is attached to the back of the EMU. It contains the oxygen tanks, cooling system equipment, radio transmitter and receiver, and batteries. There is enough oxygen and battery power in the pack to enable the astronaut to go on an eight-hour spacewalk.[7]

Fully decked out, the space shuttle EMU weighs 250 pounds. It costs over one million dollars. The good news is that the suit can be used by more than one person.[8] Before launching into space, the astronaut who will wear the suit is carefully measured, and a suit is pulled off the rack. Arms and legs are lengthened or shortened as needed. The only part of the suit that is custom-made is the gloves, because adjustable gloves are very difficult to make.

With all its layers and mechanical systems, the space shuttle EMU becomes a personal spaceship. It permits astronauts to work in the hazardous environment of space in relative safety.

4

Training for a Spacewalk

Of course, just having a space suit is not enough for a safe walk in outer space. It is important to know how to use the EMU. You can get a feel for the problem early spacewalkers faced by sitting on a swivel office chair. Try to move the chair across the room without touching your feet to the floor or grabbing anything with your hands. You soon realize that you have to be able to push on something in order to move. Astronauts need to learn how to use the handholds mounted outside the space shuttle to move in space. To do it right takes practice because it is easy to push or pull too hard and smash against the shuttle's walls.

Training for spacewalking is done in a giant swimming pool. The pool, called the neutral buoyancy laboratory (NBL), is located near NASA's Johnson Space Center in Houston, Texas. The pool is 202 feet long, 102 feet wide, and 40 feet deep. It holds 6.2 million gallons of water.[1] The pool is large enough to submerge large models of spacecraft and satellites. Astronauts training for spacewalking put on their space suits and are lowered into the water. Normally, the space suits would float in the water, but lead weights are strapped onto the suits to make them neutrally buoyant. This means that the suit floats underwater and does not rise or sink.

In the water, astronauts practice all the jobs they expect to perform in space and jobs that might be necessary in an emergency. They practice moving and using tools. The training pays off because astronauts know exactly what to expect when everything is for real. Kathryn Thornton told reporters at a press conference, "We had spent a lot of hours training in the water tank and we were ready to go do it for real."[2]

A small team of safety divers assists the spacewalkers in the tank. If a suit were to spring a leak or if the astronaut ran into some other difficulty, the safety divers could quickly bring the astronaut to the surface.

Underwater training for spacewalking has proven to be valuable many times. Years ago, a smaller version of the NBL was used to help astronauts prepare to rescue the damaged *Skylab*. *Skylab*, the first United States space station, was launched in 1973. The station was

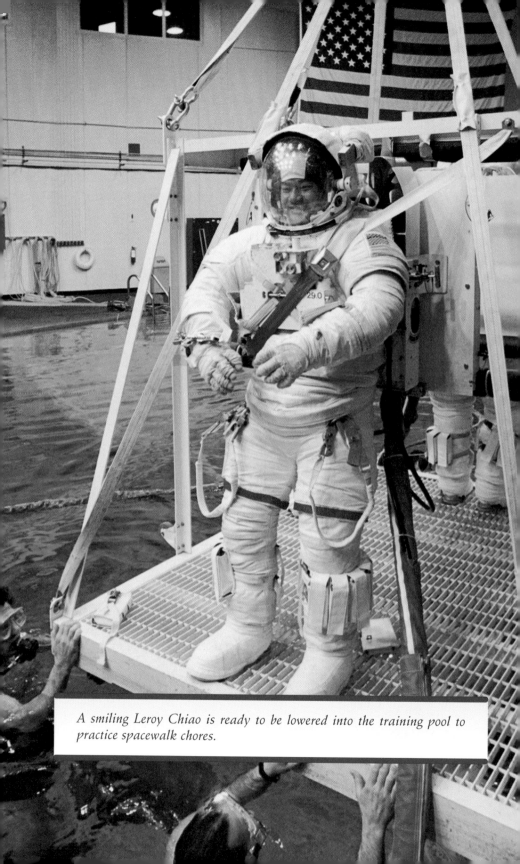

A smiling Leroy Chiao is ready to be lowered into the training pool to practice spacewalk chores.

damaged during the launch, and it needed repairs if it was going to be of any use in space.[3]

Engineers and astronauts devised plans for fixing the station. To be sure their ideas worked, astronauts put on space suits and went underwater to try them. After long hours underwater, the repair strategy was set. *Skylab* astronauts traveled to the station and were able to repair the damages during two spacewalks.[4] The job was not easy. For one task, Joseph P. Kerwin worked some long-handled cutters to snap a piece of metal holding a stuck solar panel. "Man, I am really pulling!"

"Whoops! There she goes!" hollered Kerwin's spacewalking partner, Pete Conrad.[5] The panel opened and the space station was saved.

Although the best training for spacewalking is underwater training,

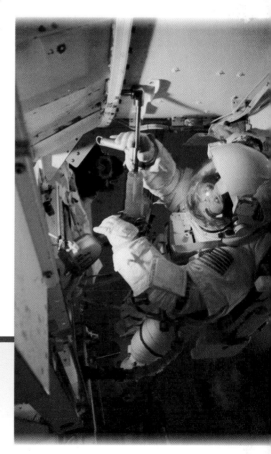

Astronaut Tamara Jernigan uses a special tool during underwater practice of a spacewalk for construction of the International Space Station.

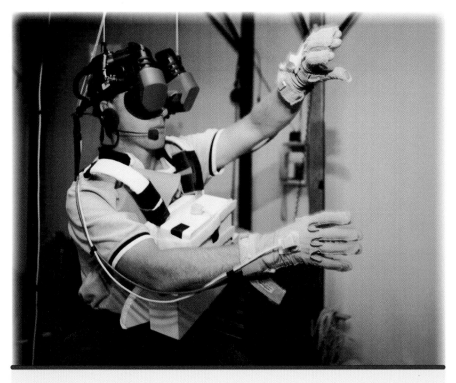

Michael Foale uses virtual reality hardware to practice his upcoming spacewalk mission.

astronauts train in other ways as well. They have standard classwork where space suit experts come and teach them about how all the parts of the space suit work. They spend long hours studying books and charts. They have many practice sessions, learning how to put on their space suits. There is training inside a mock-up of the space shuttle, which sits on the ground. The astronauts crawl into an airlock and learn how to unpack the suit and get it ready for a spacewalk. Another important job is to study the satellite they are going to repair or structure they are going to assemble. Replacing a gyroscope on the Hubble Space Telescope, for instance, required lots of knowledge.

A relatively new form of training is virtual reality. Astronauts wear helmets with small television screens for each eye. The view on the screens might be a giant space station. In this way, astronauts can study the station and get the feel of moving about it without actually being there. Virtual reality for space suit training is NASA's version of video games.

5

Time to Get Ready

A pair of astronauts getting ready for their first spacewalk realize they are in for an amazing experience. They are orbiting Earth on the space shuttle and will step out into a place that could kill them in under thirty seconds. Yet they are about to begin an adventure that may be the best time of their lives. They are about to take a spacewalk!

Astronauts start their preparations for a spacewalk by putting on oxygen masks connected to an oxygen supply line from the orbiter. The oxygen flushes out nitrogen gas from their bloodstreams. The air pressure inside their suits will be only about one third the pressure on their bodies when they are on Earth. At this lower pressure, nitrogen gas in the blood would expand into tiny bubbles in the bloodstream and cause a lot of pain. Breathing pure oxygen for thirty minutes eliminates most of the nitrogen from their blood.[1]

After they remove the oxygen masks, the astronauts start putting on their space suits. The first thing to go on is the maximum absorbency garment (MAG). The MAG is a big diaper. On top of that they wear a liquid cooling and ventilation garment. It feels good when cool water starts moving through all the tubes.

Then, they rub antifog compound inside their helmets. A bag of drinking water is placed inside each suit near the site of helmet attachment. Finally, they slip

These astronauts are breathing pure oxygen. It helps remove nitrogen from their bloodstreams before they enter the low-pressure environment of the space suit.

The black and white "Snoopy cap" is a radio headset that allows the astronaut to communicate during a spacewalk.

a "Snoopy cap" over their heads. The cap is a radio headset that makes astronauts look a bit like Snoopy in the *Peanuts* cartoon strip.

Now come the big things. They pull on the pants and move into the shuttle's airlock. This is a small room with doors on both sides. Once the inner door is closed, the door to space can be opened, and the astronauts can go out.

The top part of the suit is attached to the wall. To get in, astronauts have to push up into it from the bottom. A pair of metal rings at the waist join the top and bottom of the space suit.

Just before closing the top and bottom, spacewalkers

make a couple of electrical connections between sensors attached to their skin and the space suits' radios. These connections enable mission controllers in Houston, Texas, to monitor how well the suits are working, and enable them to keep track of the spacewalkers' heart rates and breathing rates. The radio is also used for communication.

Another astronaut helps the spacewalking partners put on the last pieces of their suits and check out the suits so that everything is perfect. The gloves come next, and then the helmets are placed over their heads and locked.[2]

The astronauts begin breathing pure oxygen again to continue eliminating nitrogen gas from their blood. To conserve the oxygen supply inside the tanks in their backpacks, a hose is connected to the front of their suits that temporarily delivers oxygen to them from the orbiter. In another forty minutes they will be ready. They go over again and again the jobs they will have to do. It is a team effort. The spacewalking partners will help each other through the jobs and assist each other if there is trouble.

The astronaut that assisted the team closes and seals the inner door to the airlock. Air pressure inside of the airlock is slowly lowered to zero to equal the pressure of space. The astronauts make leak checks to make sure their suits are sealed. If a suit is not properly sealed, the astronauts will have to fix the problem before going on.

Everything checks out. They open the round door to the payload bay. The bay is the large compartment located on the orbiter's back. Satellites and space station components (payload) are carried here. Slowly, the astronauts enter the payload bay—they are now in outer space. As astronaut Jeff Hoffman put it, "It's an exciting moment when you open the airlock and look out and there is the entire universe staring you in the face."[3]

The only sounds they hear are the noises coming from fans blowing air inside their suits and from the radio chatter. Continents, oceans, and clouds pass silently hundreds of miles beneath them as the space shuttle continues on its mission.

Astronauts might be tempted to drift out into the payload of the space shuttle, but that would be a mistake. Moving about in space is tricky. Since they are floating, they have to push on something when they want to go the opposite way. Once they are moving, they keep on moving until they can push on something else to stop themselves or change their direction. If spacewalkers were to accidentally drift above the space shuttle's payload bay where there is not anything to push on, they would continue to drift away into space. That could be a real problem. The commander of the space shuttle would have to fire the shuttle's small maneuvering rocket engines to chase after them.

The first thing astronauts do when entering the

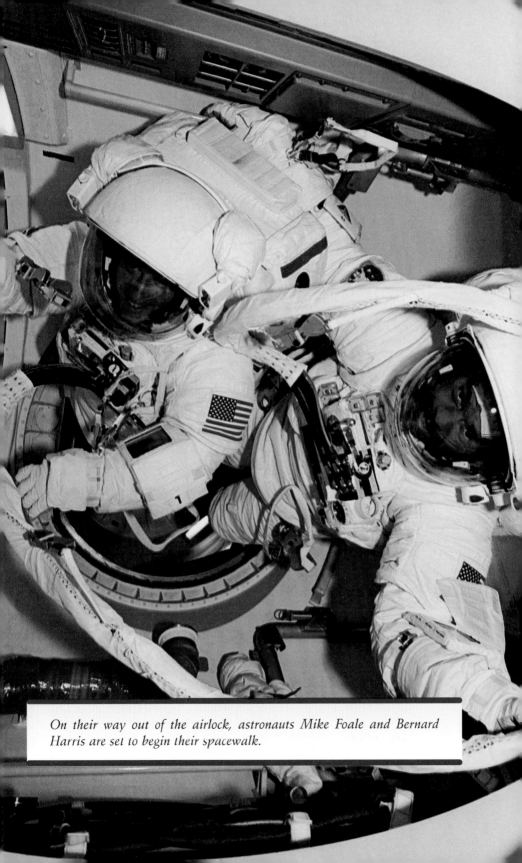

On their way out of the airlock, astronauts Mike Foale and Bernard Harris are set to begin their spacewalk.

payload bay is to clip a tether to a metal bar that serves as a handhold. The tether is a short leash that is connected to the space suit. The other end has a large clip that looks as if it would be used for a giant dog leash. The tether prevents spacewalkers from drifting away from the handholds if they let go with their hands.

Astronauts move from one handhold to another, using their tether for safety. They begin to move toward their work site. If they were working on the International Space Station, they might need to reach the docking structure of the station.

Astronauts can turn on a pair of headlights attached to their helmets as the space shuttle orbits around the dark side of Earth. There are also lights in the payload bay, but the headlights help them see better. Every forty-five minutes the shuttle passes from light to darkness and back to light. Because spacewalkers are so busy, Mission Control radios reminders to them, like the ones the STS-61 spacewalkers received. "OK, we're going over the northwest corner of Australia. . . . Visor is required. It is daylight."[4] Mission Control also sends reminders to turn on the headlights.

Astronauts carry different sets of tools with them, depending on the job they will be performing. The tools look like those found in any toolbox on Earth except for a few differences. The tools have rings to attach them to small tethers so they cannot be dropped in space. A lost wrench could be a disaster for some future spacecraft

James Newman waves to the camera during his spacewalk. Notice that he is holding a handrail on the outside of the ISS.

that collided with it in orbit. The tools also have thicker handles to make them easier to grasp with bulky space suit gloves.[5]

Astronauts on a spacewalk might use a wrench to tighten the electrical connectors of a new module they are joining on the space station. Turning nuts with a wrench is tricky. An astronaut has to anchor himself to the space shuttle so that if the nut is stuck, he does not turn himself instead of the nut.

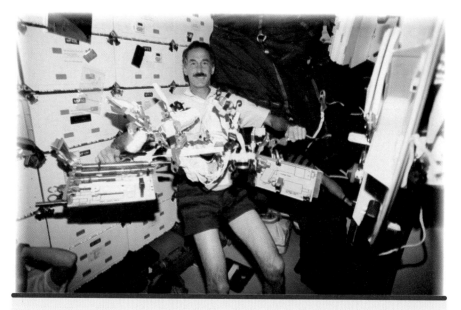

Jeffrey Hoffman displays tools used during a spacewalk repair mission.

Working closely as a team, the spacewalkers finish their jobs and start heading back toward the shuttle's airlock. They reenter the airlock and seal the door. The airlock fills with air and the inner door is opened. At this point, it is safe for the astronauts to take off their suits. With the suits stowed, the astronauts can relax and enjoy a well-deserved meal and share their experiences with their fellow astronauts.

6

Spacewalking Past and Future

The most dramatic and exciting spacewalks took place between the years 1969 and 1972. Six times, teams of three astronauts left Earth on the mighty Saturn V rocket and landed on the Moon. While one astronaut remained in the spacecraft that would return the teams to Earth, two astronauts descended to the Moon's surface in a special four-legged rocket landing craft. Climbing down a ladder on one of the craft's legs, these astronauts stepped out onto the gray, dusty surface of Earth's nearest neighbor in space.

During the six Moonwalks, astronauts collected a total of 843 pounds of lunar rock and soil for scientists to study.[1] They set up automatic science experiments that would run for years after they left. Finally, they

Commander of Apollo 16, John Young, performed one of America's most exciting spacewalks—on the Moon. He salutes the American flag while leaping from the lunar surface.

learned how to move about on the surface of the Moon where the gravity is only one sixth that of Earth. On Earth their heavy lunar space suits weighed 180 pounds, but on the Moon, they weighed only 30 pounds.[2]

Their space suits for the Moon contained backpacks with oxygen supplies like the space shuttle space suits do today. They were free to move about without any oxygen lines to drag behind them. At the time, their suits were the best ever made for spacewalking, but they were somewhat stiff and awkward to move about in. Still, the astronauts were able to use tools, set up equipment, and even drive a lunar buggy. The buggy, called the lunar rover, was an electric car that enabled the astronauts to travel long distances from their lander without getting tired. The astronauts were able to reach large craters, valleys, and giant boulders and bring back their samples in the "trunk."

It has been over thirty years since the first of the six teams of astronauts stepped on the Moon, and none have gone back. Spacewalking has now concentrated on

The lunar rover allowed astronauts to travel across the Moon's surface without getting tired.

During construction of the ISS, astronaut Tamara Jernigan is anchored to a foot restraint on the remote manipulator system. Her seven-hour spacewalk was just one of many that will be completed to build the space station.

operating outside the space shuttle in low Earth orbit. That will change soon.

Construction of the International Space Station (ISS) is under way. When completed, the station will consist of many parts, such as large cylinders for living quarters and laboratories, a long truss beam, solar panels for making electricity, and docking ports for space vehicles that come to visit. The station, which will be as large as two football fields side by side, is a combined project of the United States and fifteen other nations.[3] It has to be carried into space in many pieces and joined.

As the ISS is constructed, many astronauts will have to put on their space suits and go outside the station to connect the parts. Astronauts will do things that seem superhuman. As Jeff Hoffman was handling a massive instrument on the STS-61 mission, he found he could move it about by his fingertips: "I am not even pulling on it. I am coaxing it with my fingertips. . . . You can do things in zero-g that you would never have dreamt of."[4]

Astronauts Tamara Jernigan and Dan Barry performed a spacewalk to work on the ISS construction. They placed two cranes from the shuttle on the ISS. They also added some foot restraints and handrails to the station that future spacewalkers can use during continued assembly of the ISS.[5] By the time the ISS is finished, astronauts will log more than nine hundred hours spacewalking, doing extraordinary things.[6]

CHAPTER NOTES

Chapter 1. The Ultimate Adventure

1. STS-61 Mission Highlights Resource Tape, Reference Master 605892 (Houston: National Aeronautics and Space Administration, 1994).

2. *STS-61 Mission Highlights* (Houston: National Aeronautics and Space Administration, 1994), p. 3.

3. STS-61 Post Flight Press Conference (Houston: National Aeronautics and Space Administration, January 4, 1994).

4. STS-61 Mission Highlights Resource Tape.

5. Ibid.

6. Reporter's Space Flight Note Pad (Downey, Calif.: Boeing Reusable Space Systems Office of Communication), p. Y-46.

7. STS-61 Mission Highlights Resource Tape.

8. Reporter's Space Flight Note Pad, p. Y-46.

Chapter 2. The Hazards of Space

1. Gregory Vogt, *Suited for Spacewalking* (Washington, D.C.: National Aeronautics and Space Administration, 1998), p. 10.

2. Lillian D. Kozloski, *U.S. Space Gear: Outfitting the Astronaut* (Washington, D.C.: Smithsonian Institution Press, 1994), p. 8.

3. Frank Kuznik, "Spacesuit Saga: A Story in Many Parts," *Smithsonian, Air and Space*, vol. 12, no. 3, 1997, p. 42.

4. Marsh C. Cuttino, "Micro meteoroid orbital debris and traumatic injury during EVA," *Emergency Medicine*, Medical College of Virginia, 1998, p. 1.

Chapter 3. Building a Space Suit

1. *Above and Beyond: The Encyclopedia of Aviation and Space Sciences* (Chicago: New Horizons Publishers, Inc., 1968), vol. 7, p. 1333.

2. David Portree and Robert Trevino, *Walking to Olympus: An EVA Chronology* (Washington, D.C.: NASA History Office, 1997), p. 2.

3. Lillian D. Kozloski, *U.S. Space Gear: Outfitting the Astronaut* (Washington, D.C.: Smithsonian Institution Press, 1994), pp. 64–65.

4. Virgil "Gus" Grissom, *Gemini, A Personal Account of Man's Venture into Space* (New York: The Macmillan Company, 1968), p. 129.

5. Kozloski, p. 14.

6. Ibid., p. 25.

7. Portree and Trevino, p. 89.

8. Kozloski, p. 125.

Chapter 4. Training for a Spacewalk

1. National Aeronautics and Space Administration, *Neutral Buoyancy Laboratory*, <http://www-sa.jsc.nasa.gov/FCSD/FacilitiesOpsBranch/NBL/index.htm> (May 6, 1999).

2. STS-61 Post Flight Press Conference (Houston: National Aeronautics and Space Administration, January 4, 1994).

3. David Portree and Robert Trevino, *Walking to Olympus: An EVA Chronology* (Washington, D.C.: NASA History Office, 1997), pp. 31–32.

4. Ibid., p. 32.

5. Thomas Y. Canby, "Skylab, Outpost on the Frontier of Space," *National Geographic*, October 1974, p. 452.

Chapter 5. Time to Get Ready

1. Gregory Vogt, *Suited for Spacewalking* (Washington, D.C.: National Aeronautics and Space Administration, 1998), p. 22.

2. Ibid., p. 25.

3. STS-61 Post Flight Press Conference (Houston: National Aeronautics and Space Administration, January 4, 1994).

4. STS-61 Mission Highlights Resource Tape, Reference Master 605892 (Houston: National Aeronautics and Space Administration, 1994).

5. Vogt, p. 32.

Chapter 6. Spacewalking Past and Future

1. Gregory Vogt, *A Twenty-Fifth Anniversary Album of NASA* (New York: Franklin Watts, 1983), p. 35.

2. Lillian D. Kozloski, *U.S. Space Gear: Outfitting the Astronaut* (Washington, D.C.: Smithsonian Institution Press, 1994), p. 91.

3. "International Space Station—Overview" (Downey, Calif.: The Boeing Company, 1998), p. 1.

4. STS-61 Mission Highlights Resource Tape, Reference Master 605892 (Houston: National Aeronautics and Space Administration, 1994).

5. National Aeronautics and Space Administration, "STS-96 Extravehicular Activities," *NASA Space Shuttle*, <http://spaceflight.nasa.gov/shuttle/archives/sts-96/eva> (August 2, 1999).

6. National Aeronautics and Space Administration, "International Space Station Assembly: A Construction Site in Orbit," *NASA International Space Station*, <http://spaceflight.nasa.gov/station/eva> (August 2, 1999).

GLOSSARY

Apollo—The missions in which United States astronauts traveled to the surface of the Moon.

extravehicular activity (EVA)—Activity outside of a space vehicle (spacewalking).

extravehicular mobility unit (EMU)—Modern space suit consisting of pants, hard upper torso, display and control module, portable life support system, and helmet.

Hubble Space Telescope—Satellite that has a large telescope mirror for observing distant objects in space.

International Space Station—A large orbiting structure that sixteen nations are constructing for scientific research in space.

lunar rover—A four-wheel electric vehicle for driving on the Moon.

micrometeoroid—A small, high-speed particle from the solar system that can damage a space suit if it collides with it.

National Aeronautics and Space Administration (NASA)—Agency of the United States government that explores the atmosphere and space.

neutral buoyancy laboratory (NBL)—Large pool near the NASA Johnson Space Center used by astronauts for underwater space suit training.

payload bay—The large space in the back of the space shuttle where cargo is carried.

Saturn V—The thirty-six-story-tall rocket used to send Apollo astronauts to the Moon.

Skylab—A United States space station launched above Earth in 1973.

solar array—Large panels covered with solar cells that make electricity when sunlight falls on them.

space shuttle—The winged vehicle used by NASA for transporting astronauts and experiments to Earth orbit.

tether—A special rope, or leash, that keeps spacewalking astronauts from drifting away from their space vehicle.

FURTHER READING

Books

Berliner, Don. *Living in Space*. Minneapolis: The Lerner Publishing Group, 1993.

Cole, Michael D. *Apollo 11: First Moon Landing*. Springfield, N.J.: Enslow Publishers, Inc., 1995.

Kallen, Stuart A. *The Gemini Spacewalkers*. Minneapolis: ABDO Publishing Co., 1996.

Kennedy, Gregory. *The First Men in Space*. Broomall, Pa.: Chelsea House Publishing, 1991.

Vogt, Gregory. *Space Walking*. New York: Franklin Watts, Inc., 1987.

Internet Addresses

National Aeronautics and Space Administration. *International Space Station Homepage*. July 18, 1999. <http://spaceflight.nasa.gov/station> (August 3, 1999).

National Aeronautics and Space Administration. *Johnson Space Center Homepage*. July 9, 1999. <http://www.jsc.nasa.gov> (August 3, 1999).

National Aeronautics and Space Administration. *NASA Homepage*. August 3, 1999. <http://www.nasa.gov> (August 3, 1999).

National Aeronautics and Space Administration. *Spacelink*. n.d. <http://spacelink.msfc.nasa.gov> (August 3, 1999).

INDEX